THE COUNTRYMAN'S DIARY 1939

NOT TO BE PUBLISHED

The information given in this document is not to be communicated, either directly or indirectly, to the Press or to any person not holding an official position in HIS MAJESTY'S SERVICE.

CONTENTS

	PAGE
Burning Fuzes	3
Detonators	6
Detonating Fuzes	7
High Explosives	10
Delay Mechanisms	17
Grenades	22
Incendiaries	25
Booby Trap Mechanisms	29
Some Improvised Mines	32
Targets	34
Facts and Formulae	38
The Unit Charge	40
Safety Precautious	42

BURNING FUZES

1. SAFETY FUZE

(Bickford Fuze)

Recognition.

Safety Fuze looks like dark grey insulated wire, but instead of a metal core it contains a thin train of gunpowder. It is usually packed in black tins, containing 48 ft.

Properties.

Safety Fuze is a slow-burning fuze which will fire detonators. It burns at a rate of 2 ft. per minute, which makes it a suitable time fuze for comparatively short intervals. It will burn under water. It can be lit by matches, copper tube igniters or the cap of various mechanisms (such as the Pull Switch).

Uses.

1. It is used as the standard method of firing detonators in charges on training.
2. To provide a short delay whenever charges fired by mechanisms are made up for training purposes.

Methods of Lighting.

1. Matches. Hold head of match on to end of fuze, which should be cut on a slant, and strike *with* match box. (See Fig. 1.)

ALWAYS test a short length of fuze before using.

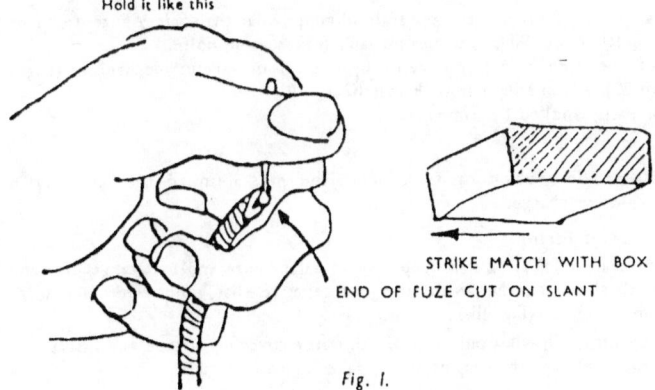

Fig. 1.

2. Copper tube Igniters. Crimp on and strike on a striker board. (See Fig. II.)

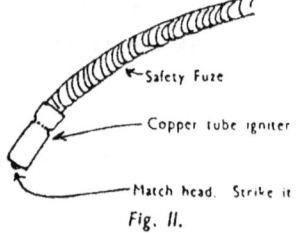

Fig. II.

3. Caps of various mechanisms. (See pages 17 and 29.)

Storage.

Keep it in taped tins in a dry place.

II. BLUE SUMP

Commercial Safety Fuze similar in all respects to Service variety except that outer covering is coloured dark blue and it is slightly less resistant to water.

III. ORANGE LINE

Recognition.

Looks like bright orange insulated wire. It is usually packed in red tins.

Properties.

Orange Line contains a thicker train of gunpowder than Safety Fuze, so burns much quicker, i.e. 90 ft. per second, and makes some noise.

It can be fired by Safety Fuze or by the cap of various mechanisms (pages 17 and 29). Don't light it with matches!

It is easily spoiled by damp.

Uses.

It is used in booby traps to connect the mechanism to the detonator in the explosive charge.

Methods of Firing.

1. By Safety Fuze. Cut a deep nick in Safety Fuze and in Orange Line and lash them together with string or tape, so that gunpowder fillings are in contact. (Fig. III.)
2. By caps. Unwind one inch of the outer covering of tape and insert into snout of cap, then crimp it to make it firm. (Fig. IV.)

Storage.

Keep it in taped tins.

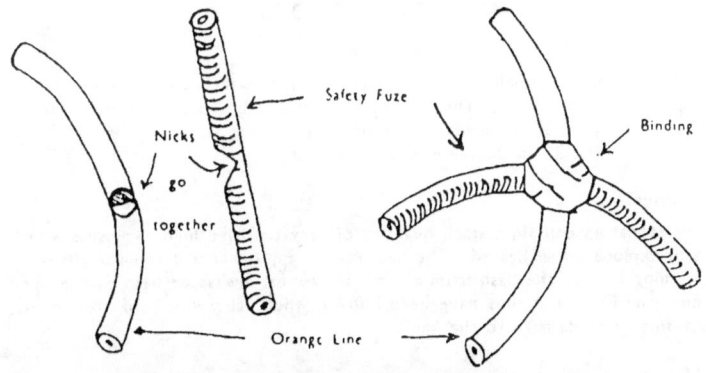

Fig. III.

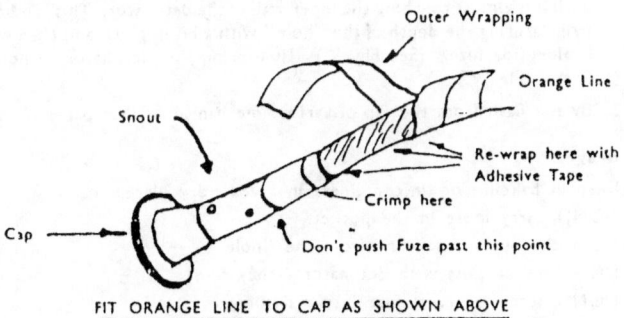

FIT ORANGE LINE TO CAP AS SHOWN ABOVE

Fig. IV.

DETONATORS

Recognition.

Detonators are small aluminium tubes about two inches long by about one-eighth of an inch wide. They are open at one end. They are usually packed
>(a) 100 in sawdust in a small tin box.
>(b) 25 in double-ended tin barrel.

Properties.

Detonators contain a small quantity of very sensitive high explosive, which will explode when heated. The heat can be applied to a detonator either by burning fuze or the flash from a cap. They must always be held by the open end (see Fig. V) as they have been known, when otherwise held, to explode, causing great damage to the hand.

Uses.

1. Detonators are used to explode charges of high explosive.
2. Detonators are used to explode Cordtex. (See page 9.)

Methods of Firing.

1. By Safety Fuze or Orange Line. Cut the end of the fuze clean and square with a sharp knife. Insert the fuze into the open end of the detonator until it is almost touching the inner wall of the detonator. This can be done by measuring the depth of the "hole" with a bit of grass, and then holding it alongside fuze. (See Fig. V.) Then crimp the detonator to hold fuze firm. (Fig. V.)
2. By the flash from the cap of various mechanisms. (See pages 17 and 29.)

Storage.

Keep in bakelite or tin containers in a cool, dry place.
DON'T carry loose in the pocket.
DO shake out any sawdust from the "hole".
DON'T be careless with detonators—they bite!
DON'T screw or twist fuze into a detonator.
Keep explosives and detonators separately in stores.

DETONATING FUZES

I. CORDTEX

Recognition.

Cordtex looks like silvery insulated wire. It is slightly thinner than Safety Fuze and has a white powder filling. It is packed usually in grey tins.

Plastic Cordtex has same filling and properties but has a white plastic cover. It is easier to handle and more reliable at high temperatures.

Properties.

Cordtex is a thin tube filled with high explosive, which can only be exploded by a detonator. The caps of various mechanisms will NOT fire it.

Though the casing is waterproof the ends are not and must be protected. The best method is to crimp on a spent .22 case. It is flexible, but bends in a circuit should not be too sharp, or it may fail. Leads of Cordtex must never cross or one of them will fail.

Uses.

1. For linking a detonator to two or more charges so that they will explode at the same instant.
2. For priming Plastic. (Fig. XII.)
3. For priming Gelignite with big knot. (Fig. XIV.)
4. For breaking A.W. Bottles in an incendiary layout.
5. As an explosive if all others run out, but only on very thin metal (e.g. petrol tanks). (Fig. VII.)

Practical Points.

1. If you want two or more charges to go off together, link them up as shown. (Fig. VI.)
2. The detonator must be firmly lashed to the Cordtex and must point towards the charges. A detonator points in the direction of its closed end. If you put a detonator on the Cordtex between two charges, it will only fire the one at which it is pointing. (Fig. VI.)

II. PRIMACORD

Recognition.

Primacord is a little thicker than Safety Fuze, and easily distinguishable by its wasp-like colouring. It is filled with a white powdery high explosive like Cordtex. Usually supplied on cardboard reels.

Properties, Uses and Storage.

Exactly as for Cordtex. Primacord contains exactly the same explosive and is interchangeable with Cordtex, for all purposes.

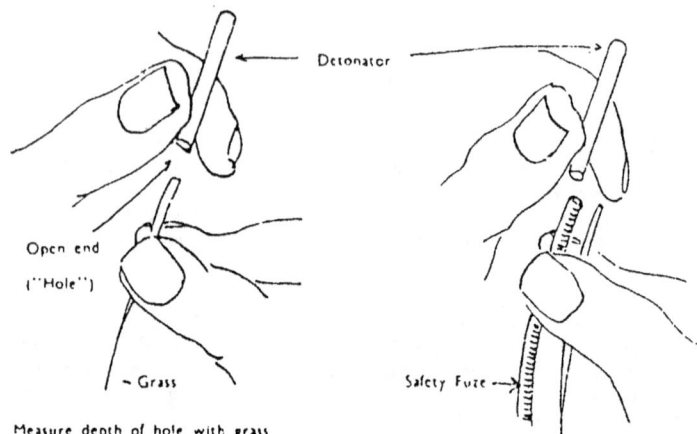

Measure depth of hole with grass

Then push fuze gently into the hole
Grass tells you how far

And crimp on the detonator to the fuze like this:

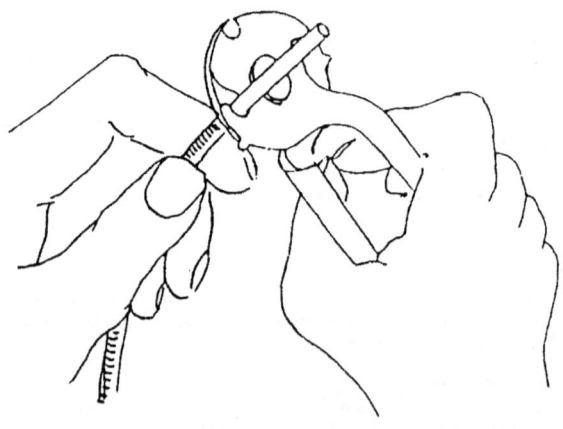

Fig. V.

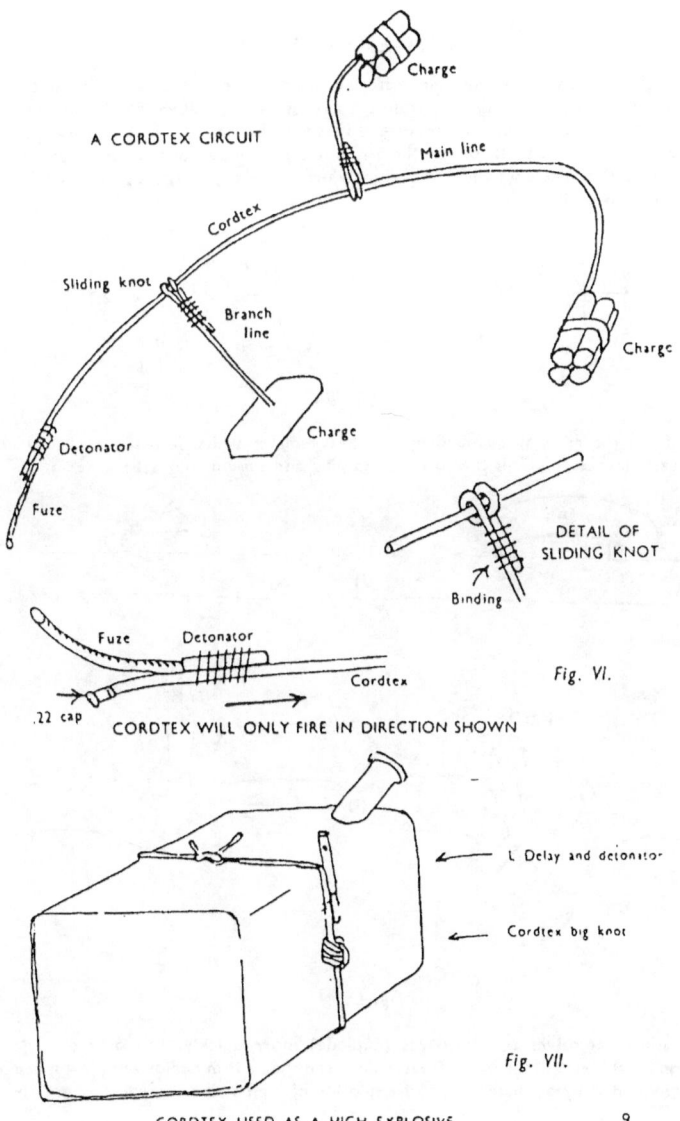

HIGH EXPLOSIVES

High explosives are unstable materials which, when suitably exploded by a detonator or primer, turn almost instantaneously into gas at very high temperature and pressure. This gas pressure pushes very hard against its surroundings. Most of this "push" is lost to the air, but a proportion of it is used against anything else in contact with the charge, which will, if the charge is great enough, be broken or bent.

Fig. VIII.

If the charge is surrounded by any heavy material, its destructive effect is greatly increased. This is known as tamping and should always be aimed at.

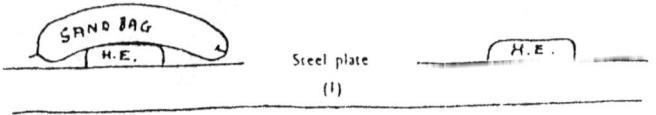

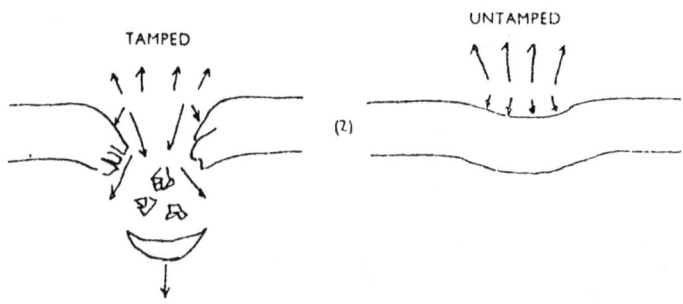

Fig. IX.

Some explosives turn into gas (explode) more quickly than others. The greater this speed, the more devastating is the effect. If an explosive of low speed is exploded by a small charge of explosive of high speed, its effect is greatly improved. This is known as priming, and the small charge as the primer. All explosives should be primed.

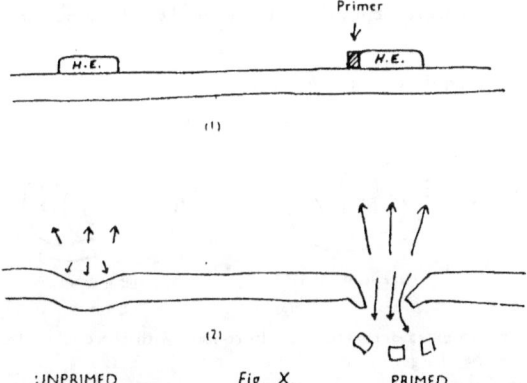

Fig. X.   UNPRIMED / PRIMED

The closer high explosive charges are laid to the object to be broken, the greater the destructive effect. This is because any air between charge and objective forms a cushion.

A charge of high explosive will always do more damage when laid in a confined space. For example:

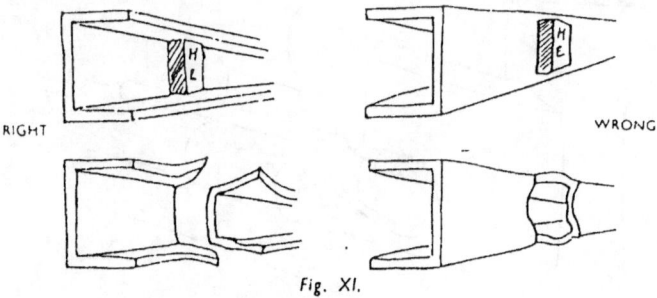

RIGHT / WRONG

Fig. XI.

I. PLASTIC

Recognition.

Plastic is a dull yellow or sometimes black substance, with the consistency of putty. It is made up in 4 ounce sausages or sticks, which are wrapped in cellophane or greaseproof paper.

Properties.

Plastic is the finest general purpose explosive in the world. It is the fastest explosive which we use, and it can be used to prime slower explosives. Owing to its being in very short supply, this is at present the only purpose for which it may be used.

It must be primed
 (i) by using a Cordtex knot. (Fig. XII.)

CORDTEX PRIMER

Make it like this   Then pull it tight

Fig. XII.

(ii) by using two extra detonators laid in contact with the one to be fired and inserted in the charge.

(iii) G.C. or C.E. primer.

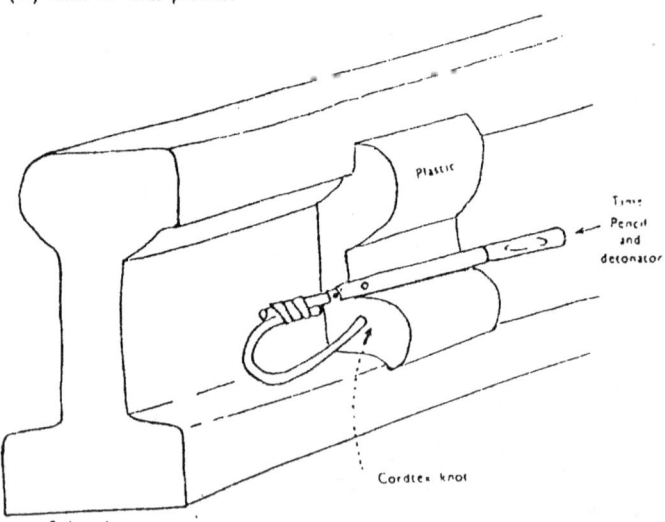

Fig. XIII.

Note.—1. Plastic thick where metal is thick.
2. Cordtex knot as primer (hidden in Plastic).
3. Time Pencil (page 17) and detonator to fire Cordtex.

Due to its putty-like consistency it can be moulded to fit objectives of any shape. It can be softened by adding vaseline or grease which is supplied in toothpaste tubes in the kit.

It is practically unaffected by storage conditions and does not deteriorate with age.

It does not cause headache.

Uses.
1. For priming other explosives.
2. For all purposes where a strong high explosive is required. (But *not* until you get some more.)

Practical Points about Using Plastic.
1. Take the required number of sticks and remove the wrapping.
2. Mould the sticks together in the hands, adding a little vaseline if necessary, until it gains a "putty-like" consistency.
3. Mould in your primer.
4. Select a weak and vulnerable part of your object and smear it with vaseline.
5. Mould on the Plastic, making it thick where the metal is thick.
6. Tape it on if necessary.
7. Tamp the charge if possible.

II. GELIGNITE

Recognition.

Gelignite is buff or greyish brown in colour, and appears to contain sawdust. It is made up in 4 oz. or 8 oz. sticks, which are always wrapped in thick buff-coloured waxed paper, marked N.S. POLAR GELIGNITE or DU PONT GELIGNITE. It is packed in waterproof paper bags or wooden boxes.

Properties.

Gelignite is not nearly so fast an explosive as Plastic, but if primed it is quite efficient. It is cheap and easy to obtain and forms our main stock of explosive at the present.

It can be exploded by a detonator alone, in which case speed may only be 2,500 metres per sec., but it is *not* efficient for cutting metal if used like this. Primed Gelignite is *four times* as powerful as unprimed. It should not be handled with bare hands, when unwrapped, as it gives a headache equal to that of a first-class hangover. As a result of this it cannot be moulded to the objective, but must be tied or wedged into position. Tape is supplied for this.

It suffers very seriously from damp storage conditions. Auxunit packing is O.K., but keep any loose sticks or cartons in taped-up tins or waterproof paper bags.

Uses.

When primed:
1. For all purposes where a cutting explosive is required.
2. For filling "made-up" charges (see pages 33 and 41).

Unprimed:
3. As a bursting charge for booby traps.

Practical Points about Using Gelignite.

1. Take the required number of sticks of Gelignite (see tables on pages 38 and 39).
2. Put them in a waterproof paper bag, so that you don't get a headache from carrying them in your pockets.
3. (a) Prime one of the sticks of Gelignite with one ounce of Plastic, which must, in turn, be primed with a Cordtex knot. Then fix the fuze and detonator to the Cordtex.
   (b) If you have no Plastic use a C.E. primer or a big knot of Cordtex as a primer. (Fig. XIV.)
4. When you reach your objective, lash either the whole bag or individual sticks to the most vulnerable part of the target with string or tape. Make certain that the sticks are touching one another, and that the primer stick is in good contact with the rest of the charge.
5. Tamp the charge if possible.

N.B.— (i) Do prime Gelignite with Plastic, or a C.E. primer, or a big Cordtex knot.
   (ii) Don't handle unwrapped Gelignite in your bare hands, or you will get a headache.

III. BLASTING GELATINE

Recognition.

Blasting Gelatine is a light buff coloured jelly. It is usually supplied in 4 oz. sticks wrapped in plain greaseproof paper. Packed in tins.

Properties, Uses, Etc.

As for Gelignite. Always prime it with Plastic, or a C.E. primer, or a big Cordtex knot.

IV. GUNCOTTON

Recognition.

Guncotton is a white solid, looking like pulp board. It is made up in 1 lb. slabs—6 in. by 3 in. by 1¼ in. in size. There is a hole in the middle of each slab.

Properties.

Guncotton is not at present supplied, but some areas have a small quantity and others may "find" some.

PRIMING GELIGNITE.

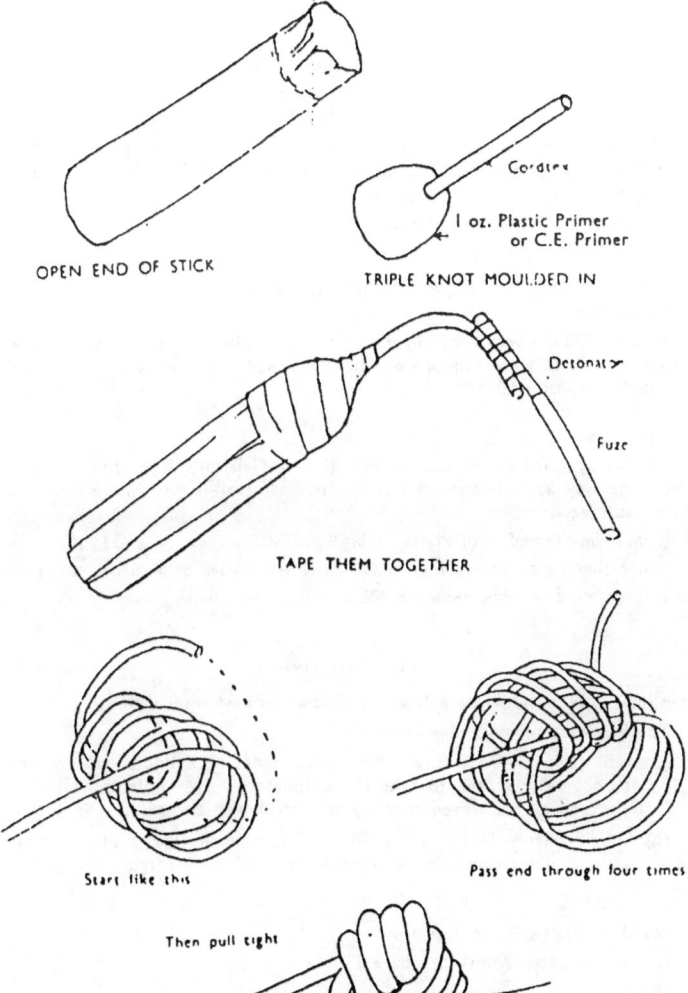

OPEN END OF STICK

Cordtex
1 oz. Plastic Primer or C.E. Primer
TRIPLE KNOT MOULDED IN

Detonator
Fuze
TAPE THEM TOGETHER

Start like this

Pass end through four times

Then pull tight

and you have

THE CORDTEX "BIG KNOT" PRIMER

Fig. XIV.

Wet Guncotton is used as a cutting charge and must be primed with a dry Guncotton primer or a C.E. primer. Providing Guncotton is kept in its air-tight packing it will not deteriorate.

IT MUST NOT BE ALLOWED TO BECOME DRY.

V. NOBEL'S EXPLOSIVE 808

Recognition.

Nobel's 808 is a very tough light buff coloured jelly, with a very strong smell of almond oil. It is made up in 4 oz. sticks wrapped in thin waxed paper marked "Nobel's Explosive 808".

Properties.

808 is a special form of desensitized Blasting Gelatine. It is used in exactly the same way as Gelignite. Its great virtue is that it does not suffer from damp storage.

It *must* be primed with Plastic, a big knot of Cordtex, or a C.E. primer.

It is rather more powerful than Gelignite when used as a cutting charge.

G.C. primers will not detonate 808 at high velocity with certainty.

VI. PRIMING.

All Explosives should be primed to obtain the best results; see Fig. X.

The following primers may be used.

(a) C.E. Primer.—This is a small waxed tapering cylinder with a hole through the centre to take the detonator. This hole must not be enlarged or the waxed covering will break and the primer be useless.

(b) G.C. Primer.—This is similar to the C.E. primer except that it is cast and the detonator hole can be enlarged without damaging the primer.

    OR

(c) 1 oz. Plastic.—See page 15.

(d) Big Cordtex Knot—See page 15.

DELAY MECHANISMS

1. THE TIME PENCIL

Recognition.

A Time Pencil looks rather like a propelling pencil. One end is copper and the other brass. They are packed in flat tins, which will open at either end.

Properties.

The coloured bands round the Time Pencil tell you how long a delay there is between the squashing of the copper tube and the firing of the cap. In the latest pattern the colour is shown on the safety strip only, hence this must not be left on the site. (See Fig. XVII.)

	Summer	Winter
Red	½ hour	¾ hour
White	1½ hours	2½ hours
Green	5 ,,	6 ,,
Yellow	10 ,,	18 ,,
Blue	20 ,,	30 ,,
Black	10 mins. (issued for training only)	

Notice the difference in delays for summer and winter. This is because the acid works more slowly when it is cold.

The cap of the Time Pencil will fire Safety Fuze, Orange Line or a detonator. It will *not* fire Cordtex.

Practical Points in Using.

Stick to the following order and you won't have any failures.

1. Select two Time Pencils of the required delay for each charge.
2. Test to see that the sight hole is clear—either by eye or with a match stick. If it is blocked, the striker has come down and the pencil is useless.
3. Examine the cap. It should be dry and pink inside. If it is brown it is damp and will not work.
4. Crimp on the detonator or fit fuze inside adaptor.
5. Squeeze the copper tube until you hear the capsule break. Don't bend it. The pencil is now set.

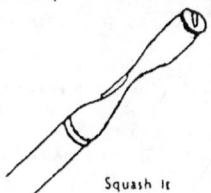
Squash it

Don't bend it

Fig. XV.

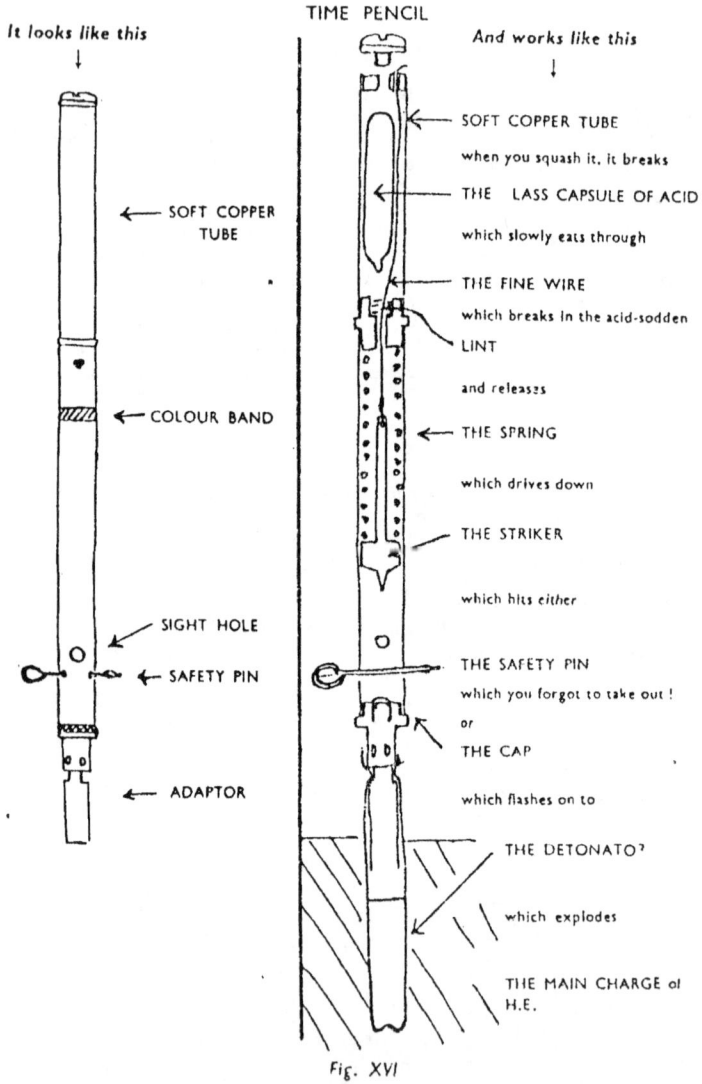

Fig. XVI

6. Test again to see that sight hole is clear. If it is blocked put the pencil aside. It is no good.
7. *Take out the safety pin.*
8. Put both Time Pencils and detonators into the charge. Fix them firmly so they won't fall out.

The reason for using two pencils is that they are not 100 per cent. reliable. By using two you are almost certain that one at least will work.

DON'T FORGET TO TAKE OUT THE SAFETY PIN.

In the most recent types the body is rather shorter, a different type of snout is fitted and the safety pin is a soft metal band which carries the colour marking, like this:—

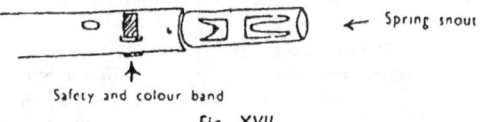

Fig. XVII.

To fix a detonator or fuze just push it into the snout, where it will be firmly held. Treat the band as a safety pin. Once it has been taken out the enemy has no way of telling the delay period.

II. "L" DELAY

Recognition.

Looks like a propelling pencil, is shorter and rather fatter than the Time Pencil. Usually wrapped in cellophane and packed in cardboard boxes of ten.

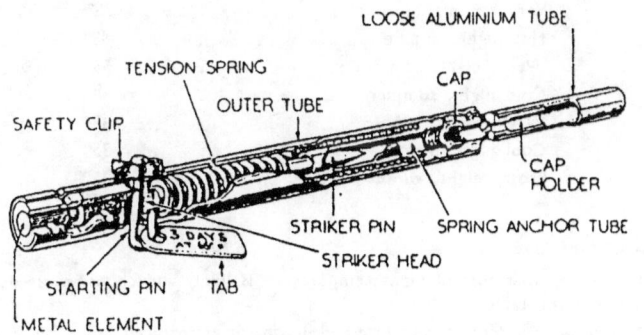

L. DELAY

Fig. XVIII.

Properties.

The "L" Delay is a new form of Time Pencil, which is used for exactly the same purposes. The great advantage is that it is less sensitive to damp and easier to use than the Time Pencil.

When the safety clip is removed it is possible to withdraw the starting pin. This frees the striker head and allows the tension spring to pull on the metal element. This slowly stretches until it breaks, so releasing the striker, which hits the cap. The flash from the cap will fire a detonator crimped on to the cap holder or snout, or light a piece of fuze crimped into the loose aluminium tube, which must also be crimped on to the snout.

The delay varies with temperature to the following scale:—

	Temperature in degrees Fahrenheit		Time in hours	
	105	¼	¾ =23 mins.	1
	75	½	1¼ =68 mins.	3
Time marked on tag	65	1	1½	4
	55	1½	2¼	
	45	2	3	8
	35	3	4½	12

It is obviously very important to estimate the average temperature. The following is a rough guide:—

Very hot night, summer	... ...	75°
Hot night, summer	... ...	65°
Warm night, summer	... ...	55°
Cold night, summer	... ...	45°
Warm night, winter	... ...	45°
Cool night, winter	... ...	35°
Frosty night, winter	... ... below	35°

Method of Use.

1. Decide what sort of night temperature is likely and choose your delay from the table.
2. Crimp on a detonator or fuze and detonator as required. (See Fig. XIX.)
3. Fit the "L" Delay to your charge.
4. Remove the safety clip.
5. Pull out the starting pin. The delay is then set. If the starting pin will not come out with a strong pull, the delay is a dud. Try another.

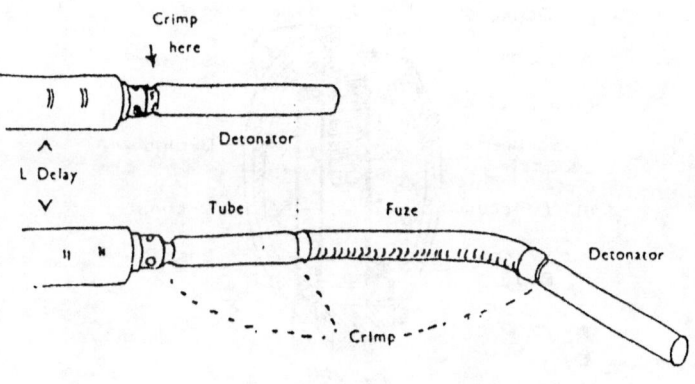

Fig. XIX.

GRENADES

1. THE 36 M OR MILLS GRENADE

Characteristics.

Weighs 1¼ lb. and can be thrown some 35 yards. It has a danger area of 20 yards in all directions from the point of burst, but fragments may have sufficient velocity to inflict wounds up to 100 yards or more, particularly if the burst is on stony ground.

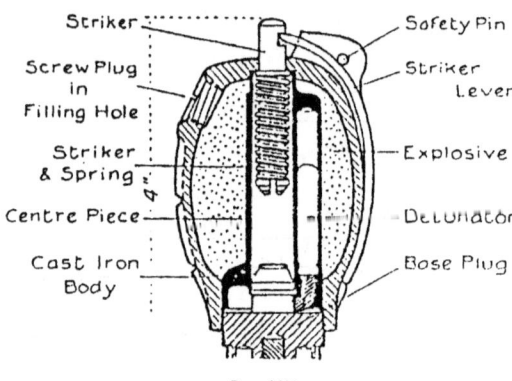

Fig. XX.

Mechanism.

Before throwing the grenade, the safety-pin must first be withdrawn. This releases the striker lever, whereupon the striker is forced down on to the cap of the igniter-set by the spring. Inside the cap there is a .22 rim fire cartridge case and a four seconds length of safety fuze.

When the striker hits the cap of the igniter-set, the cartridge case is fired and the safety fuze ignited, burning for four seconds, at the end of which time the grenade explodes.

Throwing Instruction.

Accuracy is of more importance than the distance of the throw, and any tendency to throw the grenade as far as possible without regard to accuracy should be avoided.

The grenade should be thrown at a high angle, and the best method of delivery is by an overarm swing similar to bowling in cricket.

The grenade should be held so that the striker lever is covered by the fingers only.

Before throwing, the base plug must always be removed to make sure whether the grenade is primed or not, i.e. whether the igniter-set has been fitted to the grenade. The base plug must then, of course, be replaced.

Training with Live Grenades.

(a) Training with live grenades will not take place except under the supervision of a qualified instructor.

(b) Training with live material will not take place inside any building.

(c) No smoking will take place while live material is being handled.

(d) Should a primed grenade not be expended, the igniter-set will be removed and returned to its box.

(e) Any order to take cover must be instantly obeyed.

(f) A steel helmet must be worn by every person in the party.

(g) Every "blind" will be accounted for and destroyed as follows:—

After a period of 15 minutes place a stick of primed Gelignite or Plastic, complete with detonator and fuze, in close contact with the blind, being careful not to disturb the latter. Having ascertained that everyone is under cover light the fuze and take cover yourself. After the explosion, examine whether the blind has been destroyed; if not, repeat the process.

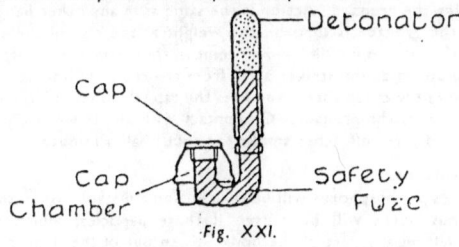

Fig. XXI.

II. THE No. 77 SMOKE GRENADE

Characteristics.

Weighs ¾ lb. and is a percussion smoke grenade producing an immediate local smoke screen. It consists essentially of a tinned plate body containing white phosphorus, a striker mechanism and a detonator.

Mechanism.

The upper portion of the grenade, normally covered by a screw-on safety cap, secured by a piece of adhesive tape, contains a ball, striker, creep spring and cap holder with flash cap. The striker is retained in position by a safety pin to which is attached a length of tape and a small lead weight. In the centre of the tin body of the grenade is a sleeve for the striker.

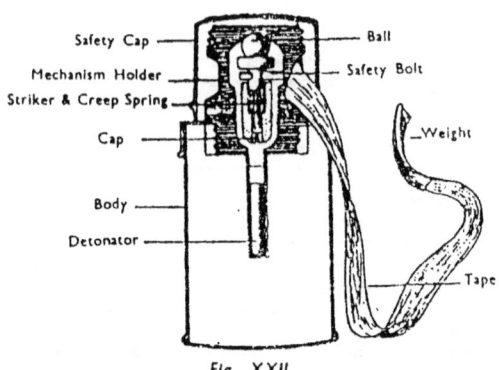

Fig. XXII.

Throwing Instructions.
  (i) To prime the grenade. Unscrew the lid at the top of the body, keeping the safety cap on. Insert a detonator (specially supplied) into the detonator sleeve, open end upwards, and replace the lid. Having removed the safety cap and keeping the safety tape in position with the fingers, the grenade is ready to throw.

  (ii) Throwing the grenade. Action is the same as in any other hand grenade. When the grenade is thrown, the weight at the end of the safety tape causes the tape to unwind and so remove the safety pin. Only the creep spring now holds the striker away from the cap. On impact, the striker overcomes the creep spring and fires the cap which sets off the detonator. This explodes the grenade. On contact with air, the white phosphorus ignites and gives off dense smoke for about half a minute.

Safety Precautions.
  Small particles of phosphorus will be thrown some distance when the grenade bursts. Serious burns will be caused if these particles, which cannot be extinguished, fall on any part of the body. Keep out of the danger zone!

  Grenades must be inspected periodically. If the body becomes corroded, air may come in contact with the phosphorus, thereby causing ignition. For the same reason, care must be taken when using the grenades on exercises or operations to see that they are not left in any position where the body may be penetrated by S.A. fire.

  Grenades should be stored away from explosives.

  Burns can be treated with a solution of copper sulphate. It is advisable to have a small quantity always available for immediate use when using these grenades.

  A solution of 2% will be sufficient.

  Blinds should be destroyed by placing a charge next to the grenade. On no account should a blind grenade be picked up.

INCENDIARIES

I. MAGNESIUM INCENDIARY

Recognition.
Black cylinder, 8 in. long and 2 in. diameter, packed in waterproof paper bags.

Properties.
The Magnesium Incendiary burns for two minutes with a very intense heat, and a large volume of sulphurous smoke. Two fuzes are fitted: one match headed for immediate use; the other plain for use with a Time Pencil. They must be kept dry.

Uses
1. In conjunction with H.E. for burning oil stores.
2. As a signal flare.

II. A.W. BOTTLE (76 GRENADE)
(also called S.I.P. or Self Igniting Phosphorus Bottle).

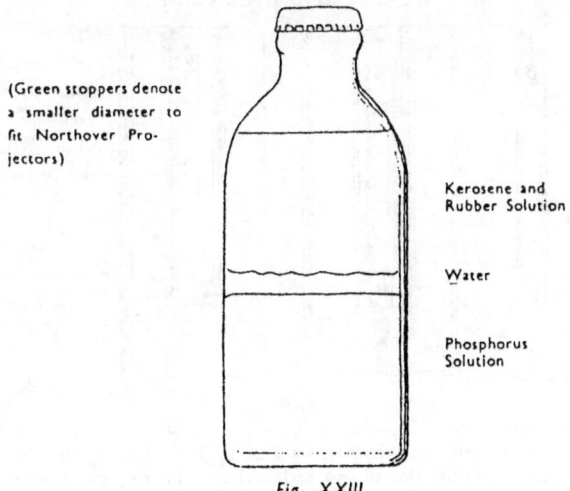

(Green stoppers denote a smaller diameter to fit Northover Projectors)

Kerosene and Rubber Solution

Water

Phosphorus Solution

Fig. XXIII.

Recognition.
Half-pint clear glass bottle filled with yellow liquids. (See Fig. XXIII.) Packed in wooden crates of 24.

Properties.

The A.W. Bottles contain yellow phosphorus and rubber dissolved in benzol. It is made of very strong glass to prevent accidental breakage.

When the bottle is broken, the contents burst into flame, and give off a dense cloud of foul-smelling smoke. The smoke is not poisonous and the service respirator is effective against it. The fire caused can be put out with water—but as soon as the water dries up the phosphorus will start burning again, and continue to do so until it is burnt right out.

Uses.

As a fire and smoke grenade in ambush of vehicles. (See below.)

Safety Precautions and Storage.

1. If a bottle is thrown and does not break, examine it very carefully for cracks before returning it to store. If cracked, destroy it.
2. Don't walk or sit upon places where these bottles have been recently thrown.
3. Store the bottles on the floor of your dump, NOT on the shelf—as they might fall off. Do not store them under water as the crown corks rust.

III. POCKET TIME INCENDIARY

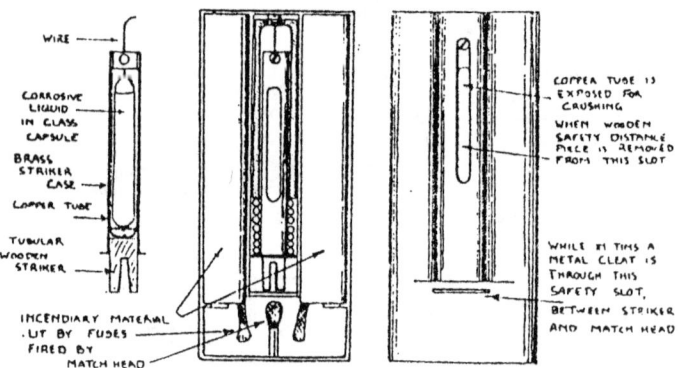

Fig. XXIV.

Recognition.

Black or mottled celluloid case 3 in. by 5 in. by ¾ in. of three tubes joined together. Supplied in flat tins of two each. The tin has two cleats which fit into the slots provided in the incendiaries and by separating the striker from the match head provide an extra safety device for transporting.

Properties.

The two outer tubes contain thermite incendiary powder candles with a burning fuze at one end. The centre tube contains a delay mechanism of the

Time Pencil variety; instead of a striker head a special forked striker sleeve is provided which, when the delay mechanism operates, flies forward and by fitting over an enlarged match head situated in the base of the incendiary ignites it. The flame produced ignites the two fuzes of the incendiary candles and this ignites the thermite.

As the celluloid case is also an incendiary material, the whole device burns extremely fiercely.

Uses.
1. As an incendiary for combustible stores.
2. Combined with an explosive charge for petrol dumps.

Method of Use.
1. Remove incendiary from metal tin.
2. Undo adhesive tape binding, thus exposing compression slot which contains a slip of wood which, until removed, acts as a safety pin.
3. Press copper tube thus exposed either with the wooden slip provided or with a coin, thus breaking glass capsule containing corrosive mixture as in Time Pencil.

Note.—The different coloured tapes round the tin boxes denote the various delays which correspond to those of the Time Pencil.

IV. FIRE POT

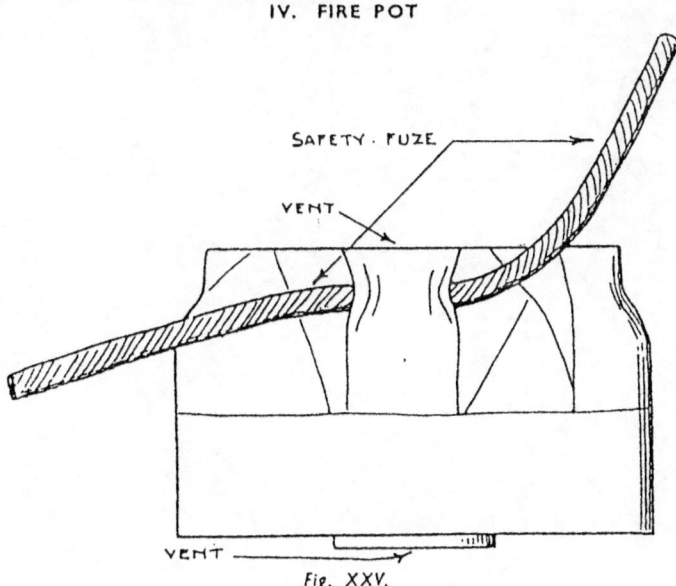

Fig. XXV.

Recognition.
Brown cylinder 2¼ in. dia. by 3 in. high.

Properties.
This Incendiary contains a very strong incendiary material which initially produces a powerful and widespread flame. After about a minute this settles down and the incendiary continues to burn more quietly for about a quarter of an hour, giving an extremely hot local flame.

Uses.
1. As an incendiary for combustible stores.
2. It is not necessary to combine it with an explosive charge as the heat is so intense that it will burst petrol tins. It is, however, recommended that a combined charge should still be used as more petrol tins will be destroyed in this way and consequently the resulting fire will be more difficult to put out.

Method of Use.
Cut two ends of Safety Fuze provided as short as possible and insert them into sleeve of Time Pencil or "L" Delay.

BOOBY TRAP MECHANISMS

I. THE PULL SWITCH

Recognition.

See Fig. XXVI. They are packed in unpainted wooden boxes of 50, marked "PULL", and in tins painted khaki also marked "PULL".

Mechanism.

The pull switch is designed so that when a wire fixed to the eye on the end is pulled, a cap is fired, which will light Orange Line, and so through a detonator fire a charge of high explosive. Here is a sectioned drawing :—

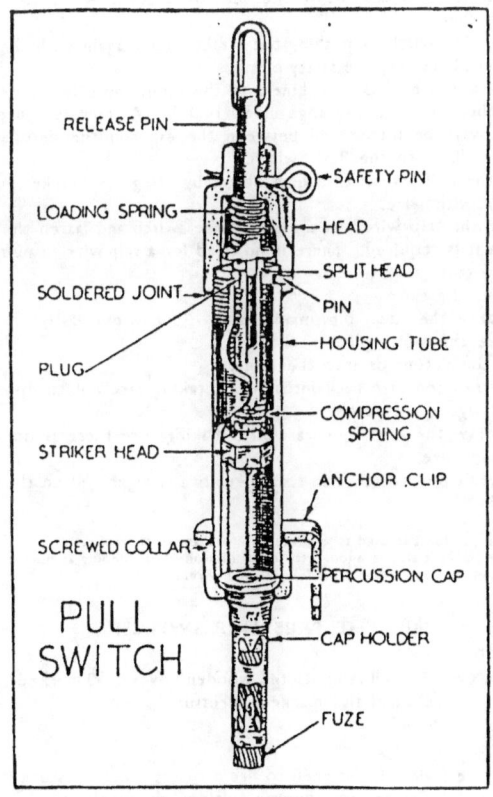

Fig. XXVI.

When a pull of about 4 lb. is applied to the release pin, the load spring is compressed and the split head of the striker allowed to contract and so pass through the plug. The striker then flies down under the influence of the compressor spring and hits the cap, which will light Orange Line.

Method of Use.
1. Test the switch. To do this—
    (a) Remove the cap.
    (b) Remove the safety pin and see that it slides freely.
    (c) Place the open end against a piece of wood.
    (d) Pull the release pin.
    The striker should fly down and dent the piece of wood. If it does not, try another.
2. Reload the switch by pushing the striker home again with the blunt end of a pencil, and replace safety pin.
3. Crimp a length of Orange Line into the snout and crimp a detonator to the other end of the Orange Line. NOTE.—A short length of Bickford will always be introduced between the cap and the detonators when TRAINING with the Pull Switch.
4. Affix the Pull Switch to a firm object by using the bracket provided, or lashing with wire.
5. Attach the trip wire* to the eye of the switch and fasten the other end where it is required. There is no need for a trip wire in a pull switch to be too taut.
6. Remove the cap.
7. Withdraw the safety pin, making sure it comes out easily.
8. Replace the safety pin.
9. Place the detonator into the mine.
10. Screw cap and fuze back into switch, taking care not to disturb it, and camouflage the switch.
11. Withdraw the safety pin carefully, making sure there is no tension on the trip wire.
12. Always fit pull switches so that there is a straight pull on the eye of the switch.

* Trap wire (.014 in.) is used to operate the mechanism of traps, etc.
Trip wire (.032) is strong enough to trip a 14 stone man running at normal speed and is used when span required is too great for the thin wire.

II. THE PRESSURE SWITCH

Recognition.

See Fig. XXVII. Packed in unpainted wooden boxes of 50 marked "Pressure" and in tins painted khaki also marked "Pressure".

Mechanism.

The Pressure Switch is designed to fire a charge when pressure is applied to the shearing pin. Facing page shows a sectional drawing.

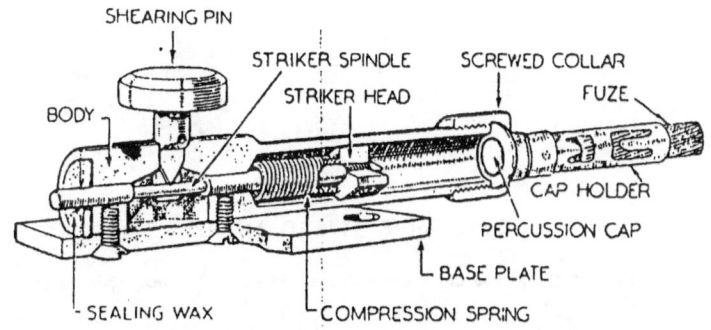

PRESSURE SWITCH

Fig. XXVII.

Pressure of about 40 lb. or more on the shearing pin snaps the striker spindle, which allows the striker head to fly forward under the influence of the compression spring and fire the cap.

Method of Use.

1. Unscrew cap and make certain that the striker is not loose. If it is, the switch is useless and cannot be reloaded.

2. Refit the cap and crimp on fuze and detonator.

3. Remove the shearing pin from the base plate and push it through the paper covering over the hole on the top of the switch, and turn it until it engages.

4. Set the switch on a firm foundation so that the load will come directly on top of the shearing pin.

5. *Remove the cap, and test the object which is to lie on top of it, to make certain that it is not too heavy.*

6. Replace the cap and insert the detonator into the mine.

7. Camouflage.

NEVER PUT THE DETONATOR INTO THE MINE BEFORE TESTING.

SOME IMPROVISED MINES

These are all things you can easily make yourselves, as the materials are easy to find.

The first step is to make a "burster". This can be made by filling a small cocoa tin with Gelignite—I lb. is enough—and moulding in a small knot of Cordtex on the end of a 2-ft. length. Bring the spare end out through the lid.

The next step is to make the "shrapnel" part of the mine. The shrapnel itself is usually small pieces of scrap metal such as nuts and bolts—nails, etc. If you can't get these use small, sharp stones. Pack these into a bigger tin around the burster. The more shrapnel the better—but remember you will have to carry the thing about.

Another idea is to bury the burster in a small heap of stones.

Another very good method is to use an old motor cycle cylinder filled with Gelignite. The fins fly very well.

Yet another is to link several 4-oz. sticks of Gelignite together by passing a length of Cordtex through the centre of each and leaving a space of 9 in. or so between them. Then put this string of sausages into a length of scrap cast iron pipe.

With a little imagination dozens of other ideas will present themselves, and they will probably be better suited to your part of the country. The essential point is that for outdoor booby traps you must aim at killing by splinters—not by blast.

(See Fig. XXVIII.)

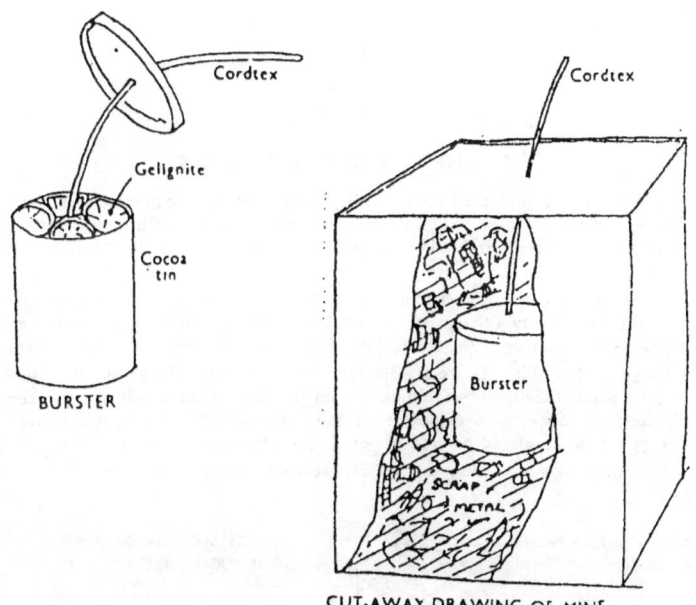

CUT-AWAY DRAWING OF MINE

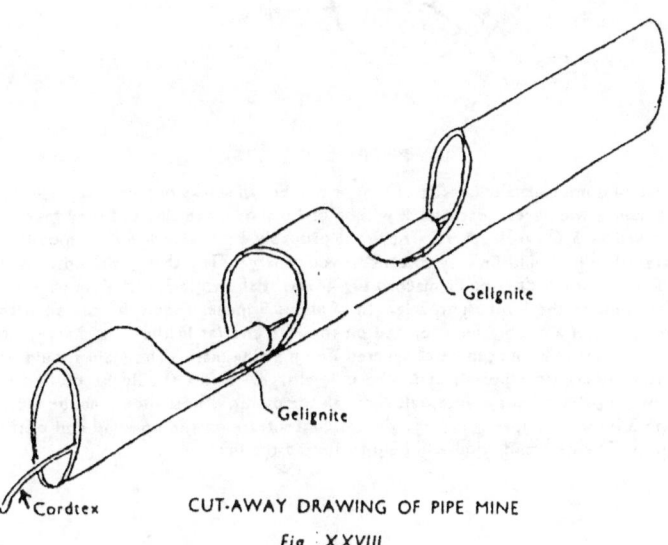

CUT-AWAY DRAWING OF PIPE MINE

Fig. XXVIII.

TARGETS

I. SHELL AND BOMB DUMPS

In the initial stages shell dumps will probably be fairly common, but bomb dumps will be left in France. In the later stages (if any) bomb dumps may be established in these islands. Dumps will always be spread out in the same way as petrol dumps.

Small shells are usually packed in wood, wicker or metal cases. If possible, open one of these and lay a charge of 1 lb. Plastic or 1¼ lb. primed Gelignite in direct contact with the side of the shell. Time Pencils can be used to fire the charges.

Large shells and bombs are less likely to be cased, and a charge of 2 lb. Plastic or 3 lb. primed Gelignite should be laid on the side. If the bombs are unfuzed, fill the fuze pocket with explosive instead. (Fig. XXIX.) If bombs are fuzed, charge should be placed on side of bomb opposite the fuze pocket.

If shells or bombs are found in a lorry, attack them—not the lorry.

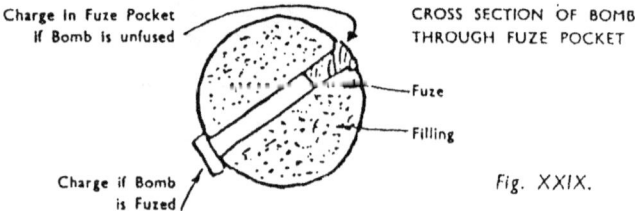

Fig. XXIX.

II. PETROL DUMPS

Petrol dumps usually consist of a large number of stacks of petrol tins spread out over a wide area. Each stack will be of 1,000 or more tins and they may be as much as 500 yards apart. They will probably be covered with camouflage material so it should be easy to conceal your charge. The charge should consist of a series of Cordtex or Primacord big knots, as shown in Fig. XIV on page 15. After pulling the knot tight, a length of about 9 inches should be cut off with the knot and a double loop formed on the free end (as in the unit charge), so that a series of knots can be connected up on a ring main. The main should be detonated by time pencils as in the unit charge. Knots should be pushed in between pairs of cans at intervals through the dump. Knots should *not* be more than 3 Jerry cans apart. In large dumps concentrate on the up-wind end of the dump. The heat and wind will help to spread the fire.

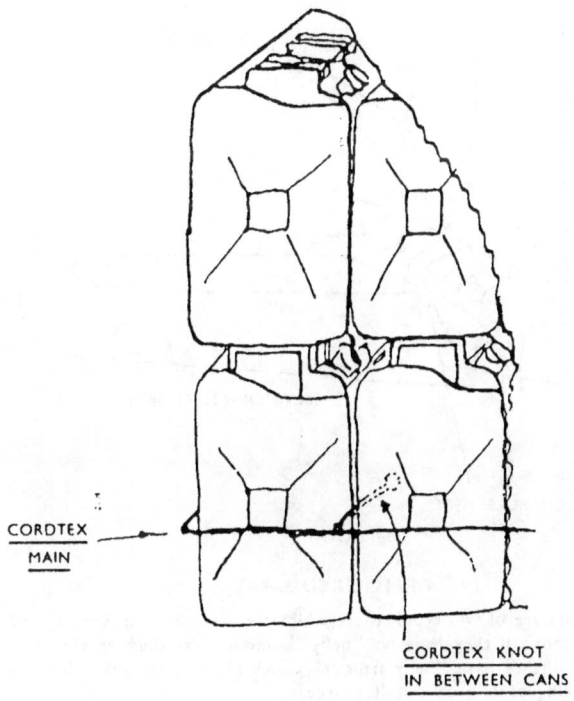

Fig. XXX.

III. AEROPLANES

The tail is the best part to attack. The following are two good methods:—

(a) Place a unit charge *inside* the fuselage at the top of the tail wheel. (There is usually an opening so that the tail wheel can be drawn up during flight.)
(b) Place a unit charge between the flat end of the elevator and the corresponding flat surface on the fuselage shown as *XX* in Fig. XXXI, or between the bottom of the rudder and the corresponding flat surface on the top of the fuselage shown as *YY* in Fig. XXXI.

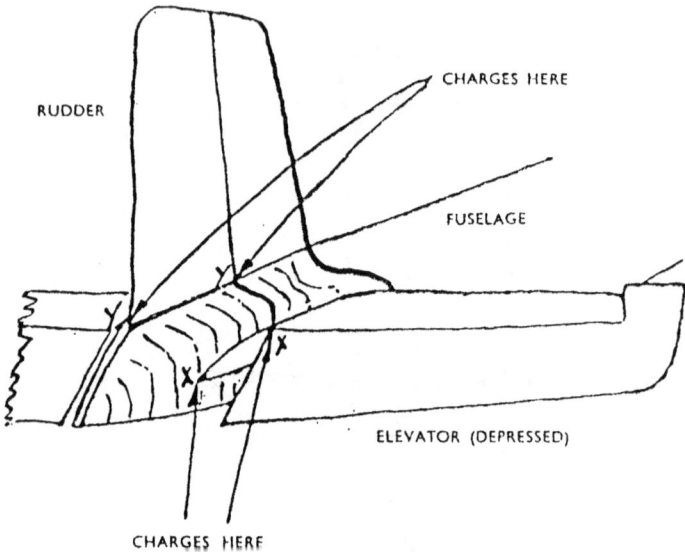

Fig. XXXI.

IV. ARMOURED CARS

Armoured cars are of two types, those which are armoured underneath and those that are not. If they have no "belly" armour treat them in the same way as lorries. If they have "belly armour" attack the steering gear, the stub axles, or the king pins, with a unit charge.

V. SEMI-TRACKED VEHICLES

1. Fix a charge of 2 lb. primed Gelignite or 808 at any one of the following points:—
    - (a) Behind or on the driving sprocket.
    - (b) Between the overlap of the idler and last bogey wheel.
    - (c) On the side of the engine.

VI. TRUCKS

1. Fix a unit charge at any of the following points:—
    - (a) Main frame at the point where the front shackle of the rear spring is attached.

(b) On the axle between the brake drum and the spring.

(c) On the differential casing.

2. Of the above, (a) and (b) are the best, as they render the vehicle incapable of being towed.

3. If time and stores permit, two of the above may be connected to fire simultaneously.

VII. TANKS

1. Fix a charge of 2 lb. of primed Gelignite or 808 behind the driving sprocket (the one with teeth on it) or against the joint of the turret and the tank.

2. Fix 1 lb. of Plastic or primed Gelignite to the outside of the smaller guns or inside the larger guns.

3. Drop 4 oz. of Plastic or primed Gelignite through any air breathers or louvres above or at the side of the engine.

4. Best of all, attack the repair lorries which always go into "harbour" with the tanks and let the field army do the "tank bursting."

NOTE.—Charges of more than $\frac{1}{4}$ lb. may be made up as one charge or by strapping the requisite number of unit charges together.

FACTS AND FORMULAE

In the formulae below the letters have the following significance:—

- C = Weight of charge in lb.
- W = Weight of charge in oz.
- D = Diameter of object to be attacked in feet.
- d = „ „ „ „ „ Inches.
- c = Circumference „ „ „ „ Inches.
- b = Width „ „ „ „ Inches (i.e. length of charge)
- B = Width „ „ „ „ feet.
- T = Thickness „ „ „ „ feet.
- t = Thickness „ „ „ „ Inches.

All formulae are for Plastic and primed explosive only.

(IF IN DOUBT, DOUBLE CHARGE AS CALCULATED.)

CUTTING CHARGES

1. **Iron and Steel Rounds including Cables** (i.e. axles, half shafts, steel cables, etc., etc.).

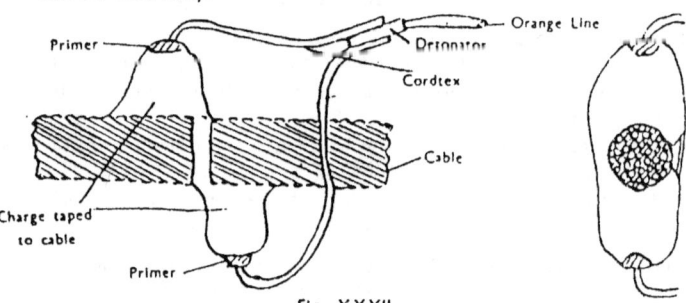

Fig. XXXII.

For cables split the charge and place half either side, so that there is no chance of cable swinging away and failing to be cut. (See Fig. XXXII.)

$$W = 4d^2 \text{ or } \frac{c^2}{2} \text{ oz.}$$

Thus for 1 in. dia. axle charge would be 4 oz.*
- 2 in. dia. „ „ „ „ 16 oz.
- 3 in. dia. „ „ „ „ 36 oz.
- 4 in. dia. „ „ „ „ 64 oz.

* Minimum charges are 2 oz. Plastic or 4 oz. of primed explosive.

And for 1 in. circumference steel cable charge would be 4 oz.*
        2 in.   ,,   ,,   ,,   ,,   ,,   ,,   4 oz.
        3 in.   ,,   ,,   ,,   ,,   ,,   ,,   9 oz.
        4 in.   ,,   ,,   ,,   ,,   ,,   ,,   16 oz.

2. Iron and Steel Plates.

$W = 3/2 bt^2$ (minimum $t = 1$)

Thus to make a cut 6 in. long in $\frac{1}{2}$ in. thick steel plate takes   4 oz.*
   ,,   ,,   ,,   6 in.   ,,   ,, 1 in.   ,,   ,,   ,,   ,,   9 oz.
   ,,   ,,   ,,   6 in.   ,,   ,, 2 in.   ,,   ,,   ,,   ,,   36 oz.
   ,,   ,,   ,,   6 in.   ,,   ,, 3 in.   ,,   ,,   ,,   ,,   81 oz.
   ,,   ,,   ,,   6 in.   ,,   ,, 4 in.   ,,   ,,   ,,   ,,   144 oz.

3. Rails.

1 lb. will cut a 90 lb. rail.
2 lb. ,, ,, 90 lb. ,, at fishplate junction.

4. Timber.

(i.) **Round Spars and Trees**

$C = 2D^2$ for soft woods.
$C = 4D^2$ for hard woods.

Thus for 6 in. tree charge would be $\frac{1}{4}$ lb. ⎫
   ,, ,, 1 ft. ,,   ,,   ,,   ,,   2 lb. ⎬ For soft woods
   ,, ,, 2 ft. ,,   ,,   ,,   ,,   8 lb. ⎭

This formula is not suitable for trees greater than $2\frac{1}{4}$ ft. diam. Trees larger than $2\frac{1}{4}$ ft. require special equipment not issued to Aux. Units.

For tree felling always place charge near foot of tree and on same side as it is required to fall. It is no use with trees already leaning in the wrong direction or in a high wind blowing in the wrong direction.

(ii.) **Timber Baulks.**

$C = 3/2 \; BT^2$

Thus for 3 in. × 3 in. charge will be   4 oz.*
   ,, ,, 4 in. × 4 in.   ,,   ,,   ,,   4 oz.*
   ,, ,, 6 in. × 6 in.   ,,   ,,   ,,   4 oz.*
   ,, ,, 9 in. × 9 in.   ,,   ,,   ,,   10 oz.
   ,, ,, 1 ft. × 1 ft.   ,,   ,,   ,,   24 oz.
   ,, ,, 2 ft. × 2 ft.   ,,   ,,   ,,   12 lb.

All cutting charges should be made up to give the maximum amount of explosives to each sq. inch of target, i.e. charges should be placed on edge rather than on the flat; this makes the difference between a chisel cut and a hammer blow.

Good contact with the surface to be cut is essential and this can be achieved by flattening out the cartridge of explosive in contact with the target.

If possible, "tamp" all cutting charges, i.e. make it as difficult as possible for the gases to escape; this can be done by covering the charge with damp clay and/or placing it in as confined a space as possible.

When using primed explosive always have the primed stick on the outside away from the target.

* Minimum charges are 2 oz. Plastic or 4 oz. of primed explosive.

THE UNIT CHARGE

The ½ lb. charge shown in Fig. XXXIII provides one that can be used on most targets, such as vehicles, dumps, guns, etc. It can be made up before setting out and some damage can always be done with it, even if the type of target is not known beforehand.

The Charge.

Is made of one stick of 8 oz. or two sticks each of 4 oz. of primed Gelignite, Gelatine or 808 or Plastic.

One stick only is primed.

A length of Cordtex or Primacord is moulded into the Plastic primer, or knotted through a C.E. primer, and to this two Time Pencils with detonators attached are taped. At the other end a double loop is made in the Cordtex (or Primacord).

You can strap magnets on as well if you like.

Carried separately are two or three lengths of Cordtex (or Primacord) with two Time Pencils with detonators attached taped to one end of each. These lengths should vary from 10 to 30 ft.

Carriage.

The charges should be carried inside the battledress blouse where 9 or 10 can be carried by one man. They must be carried in a waterproof paper bag or they will give you a headache.

The lengths of Cordtex (or Primacord) round the waist.

To Use.

Where only one charge is sufficient, it is placed and the two Time Pencils squeezed.

Where more than one charge is required, any number can be placed and connected up by passing a length of Cordtex (or Primacord), as described above, through each loop and squeezing the Time Pencils on the Cordtex (or Primacord). **DON'T SQUEEZE THE PENCILS ON THE CHARGES. DON'T FORGET THE SAFETY PINS.**

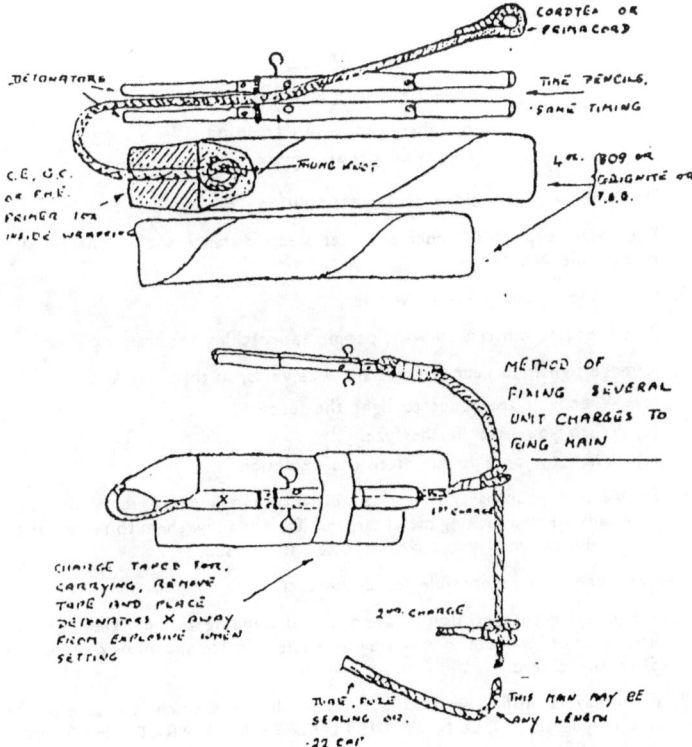

THE ½ LB UNIT CHARGE

Fig. XXXIII.

GENERAL
SAFETY PRECAUTIONS
to be observed when training
WITH EXPLOSIVES

1. Don't allow any smoking during Demolition Training.

2. Keep your explosives under cover at a safe distance and in charge of a responsible N.C.O.

3. Don't allow your party to wander.

4. Post adequate sentries to warn people approaching the training area.

5. Arrange signals to your sentries and vice versa, so that they know:—
   (a) When you are about to light the fuze.
   (b) When you have lit the fuze.
   (c) When the area is safe after the explosion.

6. Ensure that your party is at a safe distance and under cover if possible, especially when attacking metal targets. Do not allow them to return after the explosion until you have inspected it yourself.

7. Make one man responsible for detonators.

8. When laying charges, don't have a crowd standing round. Demonstrate first, then order them to withdraw to a safe distance and allow one man to place the charge.

9. If you have a misfire, wait 30 minutes and lay another charge close to the misfire yourself. **DO NOT DISTURB THE CHARGE THAT HAS MISFIRED.**

10. **DON'T TAKE UNNECESSARY RISKS.**

www.ingramcontent.com/pod-product-compliance
Lightning Source LLC
Chambersburg PA
CBHW050247230526
45470CB00005B/2151